Percent Applications

Allan D. Suter

McGraw Hill Contemporary

D0898856

Series Editor: Mitch Rosin
Executive Editor: Linda Kwil
Production Manager: Genevieve Kelley
Marketing Manager: Sean Klunder
Cover Design: Steve Strauss, ¡Think! Design

 **Contemporary**

Send all inquiries to:
McGraw-Hill/Contemporary
130 East Randolph Street, Suite 400
Chicago, Illinois 60601

ISBN: 0-07-287113-X

Printed in the United States of America.

1 2 3 4 5 6 7 8 9 10 QPD/QPD 09 08 07 06 05 04 03

The **McGraw·Hill** Companies

■ Contents

1. In the statement "50% of 14 is 7," which number is the percent?

   Answer: _____

2. In the statement "25% of 80 is 20," what number represents the part?

   Answer: _____

3. In the statement "$33\frac{1}{3}$% of 27 is 9," what is the total?

   Answer: _____

4. Find 60% of 75.

   Answer: _____

5. 200% of 154 is what number?

   Answer: _____

6. What is 25% of 55?

   Answer: _____

7. 15% of 420 is what number?

   Answer: _____

8. What is $12\frac{1}{2}$% of 320?

   Answer: _____

9. 13 is what percent of 50?

   Answer: _____

10. What percent of 12 is 9?

    Answer: _____

11. 12 is what percent of 32?

Answer: _____

12. What percent of 27 is 18?

Answer: _____

13. What percent of 30 is 45?

Answer: _____

14. 25% of what number is 15?

Answer: _____

15. 60% of what number is 90?

Answer: _____

16. Paolo bought a ten-speed bike for 25% off the original price of $390.00. How much did he save?

Answer: _____

17. On a test of 20 problems, Nina had 17 correct answers. What percent of the problems did she have correct?

Answer: _____

18. Mr. Woods earns a 5% commission on total sales each week. Last week he earned a $95.00 commission. How much were his total sales?

Answer: _____

19. Marla bought a coat that originally sold for $120. She saved $18 buying the coat on sale. The amount she saved was what percent of the original price?

Answer: _____

20. On a snowy evening, 6% of 350 registered night-school students were absent. How many students were absent that evening?

Answer: _____

Evaluation Chart

On the following chart, circle the number of any problem you missed. The column after the problem number tells you the pages where those problems are taught. Based on your score, your teacher may ask you to study specific sections of this book. However, to thoroughly review your skills, begin with Unit 1 on page 7.

Skill Area	Pretest Problem Number	Skill Section	Review Page
Percent Problems	All	7–11	12
Find the Part	2, 4, 5, 6, 7, 8	13–21	22
Find the Percent	1, 9, 10, 11, 12, 13	7–11 23–35	12 36
Find the Total	3, 14, 15	7–11 37–42	12 43
Percent Problem Solving	16, 17, 18, 19, 20	44–55	56
Life-Skills Math	All	57–72	73, 74

Percent Applications

We use percents in many different situations every day. Percents are used in business, sports, budgets, and taxes. They are helpful in solving everyday problems and in showing relationships between numbers.

A *percent* is a *part of 100.*

25%off

Regular price
$96

Save $24

Number sentence: 25% of $96 = $24

There are 3 important items to consider.

Item 1: **Total** ⟶ $96 is the **total**.

Item 2: **Part** ⟶ $24 is the **part** of the total.

Item 3: **Percent** ⟶ 25% is the **percent**.

30%off

Regular price
$5.00

Save $1.50

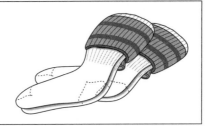

Number sentence: 30% of $5.00 = $1.50

1. In the number sentence above,
   what price represents the total? _____

2. What price (part) is being compared to the total? _____

3. What percent is given? _____

Percent Readiness

percent	total	part

25% of 4 is 1

Study each number sentence and diagram. Name the part, total, and percent for each.

Number Sentence	Diagram	Part Circled	Total—100% of Xs	Percent
1. 50% of 4 is 2.	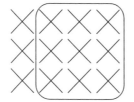	2	4	50%
2. 75% of 12 is 9.	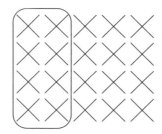	_____	_____	_____
3. 40% of 20 is 8.		_____	_____	_____
4. $33\frac{1}{3}$% of 9 is 3.		_____	_____	_____
5. 20% of 10 is 2.		_____	_____	_____
6. $62\frac{1}{2}$% of 8 is 5.		_____	_____	_____

Understanding Number Sentences

Draw a diagram for each number sentence using Xs. Then name the part, total, and percent.

Number Sentence	Diagram Each Number Sentence Using Xs	Part Circled	Total— 100% of Xs	Percent
1. 25% of 8 is 2.	$\bigotimes\bigotimes\bigotimes\bigotimes$ $\bigotimes\bigotimes\bigotimes\bigotimes$ Diagram	2	8	25%
2. 75% of 4 is 3.	Diagram	_____	_____	_____
3. 20% of 15 is 3.	Diagram	_____	_____	_____
4. $33\frac{1}{3}$% of 3 is 1.	Diagram	_____	_____	_____
5. 30% of 10 is 3.	Diagram	_____	_____	_____
6. $12\frac{1}{2}$% of 16 is 2.	Diagram	_____	_____	_____

Showing Relationships

Number Sentence: 40% of 10 is 4.

Answer the following questions about the number sentence above.

1. What number represents the total, or 100%? _____

2. What number represents the part? _____

3. What percent is given? _____

4. Draw a picture inside the box below using Xs to represent this number sentence: 40% of 10 is 4.

Circle the part.

Fill in the chart.

	Number Sentence	Total	Part	Percent
5.	10% of 70 is 7.	70		
6.	60% of 50 is 30.		30	
7.	30% of 10 is 3.			30%
8.	$66\frac{2}{3}\%$ of 9 is 6.			
9.	$33\frac{1}{3}\%$ of 300 is 100.			
10.	50% of 90 is 45.			
11.	$83\frac{1}{3}\%$ of 12 is 10.			
12.	70% of 60 is 42.			

Percent Applications

Identify the Part, Total, and Percent

For each number sentence, write the part, the total, and the percent.

1. 16 is 20% of 80.   _____ ___80___ _____
                        part   total   percent

2. 70% of 30 is 21.   ___21___   _____ _____
                        part   total   percent

3. 5.7 is 15% of 38. _____ _____ ___15%___
                        part   total   percent

4. 200% of 96 is 192. _____ _____ _____
                        part   total   percent

5. 8.04 is 12% of 67. _____ _____ _____
                        part   total   percent

6. 90% of 70 is 63.   _____ _____ _____
                        part   total   percent

7. 267 is 300% of 89. _____ _____ _____
                        part   total   percent

8. 40% of 70 is 28.   _____ _____ _____
                        part   total   percent

Percent Problems Review

For each problem, name the part, the total, and the percent.

1. 25% of 12 is 3.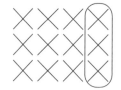

   a) Part: _____

   b) Total: _____

   c) Percent: _____

2. 40% of 10 is 4.

   a) Part: _____

   b) Total: _____

   c) Percent: _____

3. 50% of 8 is 4.

   a) Part: _____

   b) Total: _____

   c) Percent: _____

4. $33\frac{1}{3}$% of 6 is 2.

   a) Part: _____

   b) Total: _____

   c) Percent: _____

5. 200% of 45 is 90.

   a) Part: _____

   b) Total: _____

   c) Percent: _____

6. 10.6 is 25% of 42.4.

   a) Part: _____

   b) Total: _____

   c) Percent: _____

7. 259 is 350% of 74.

   a) Part: _____

   b) Total: _____

   c) Percent: _____

8. 30% of 90 is 27.

   a) Part: _____

   b) Total: _____

   c) Percent: _____

Find the Percent of a Number

SALE
10%
off

Regular price
$120

How much do you save if you buy the bicycle on sale?

To find the part of a number, change the percent to a decimal and multiply.

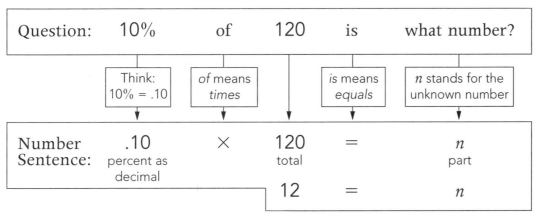

| Question: | 10% | of | 120 | is | what number? |

| | Think: 10% = .10 | *of* means *times* | | *is* means *equals* | *n* stands for the unknown number |

Number Sentence:	.10	×	120	=	*n*
	percent as decimal		total		part
			12	=	*n*

You save $12 by buying the bicycle on sale.

> Remember: To change a percent to a decimal, move the decimal point two places to the left.

$$30\% = 30. = .30 \qquad 5\% = 05. = .05 \qquad 120\% = 120. = 1.20$$

Change the percent to a decimal and multiply.

1. 30% of 70 is what number?

   _____ × _____ = ___*n*___
   percent as   total   part
   a decimal

   _____ = *n*

2. 80% of 50 is what number?

   _____ × _____ = ___*n*___
   percent as   total   part
   a decimal

   _____ = *n*

Change Percents to Decimals

For each question:

1. Change the percent to a decimal.

2. Write a number sentence.

3. Solve for n (part).

EXAMPLE

33% of 55 is what number?

$33\% = .33$

$.33 \times 55 = n$

$$
\begin{array}{r}
55 \\
\times\ .33 \\
\hline
165 \\
1650 \\
\hline
18.15
\end{array}
$$

← 2 decimal places

← 2 decimal places

---

1. 19% of 17 is what number?

_____ × _____ = _____
percent as   total   part
a decimal

_____ = n

4. 82% of 194 is what number?

_____ × _____ = _____
percent as   total   part
a decimal

_____ = n

---

2. 95% of 960 is what number?

_____ × _____ = _____
percent as   total   part
a decimal

_____ = n

5. 48% of 39 is what number?

_____ × _____ = _____
percent as   total   part
a decimal

_____ = n

---

3. 8% of 23 is what number?

_____ × _____ = _____
percent as   total   part
a decimal

_____ = n

6. 15% of 598 is what number?

_____ × _____ = _____
percent as   total   part
a decimal

_____ = n

Small and Large Percents of a Number

To find more than 100% of a number:	To find less than 1% of a number:
• Change the percent to a decimal.	• Change the percent to a decimal.
• Multiply.	• Multiply.

A. 127% of 89 is what number?

Think: 127% = 1.27 = 1.27

$1.27 \times 89 = n$

$\underline{} = n$

fill in

B. .4% of 89 is what number?

Think: .4% = .00.4 = .004

$.004 \times 89 = n$

$\underline{} = n$

fill in

Find the part. First change the percent to a decimal.

1. 200% of 348 is what number?

   Think: 200% = 2.00

   $2 \times 348 = n$

   $\underline{} = n$

4. .8% of 32 is what number?

   Think: .8% = .008

   $.008 \times 32 = n$

   $\underline{} = n$

---

2. 112% of 95 is what number?

5. .3% of 285 is what number?

---

3. 345% of 70 is what number?

6. 2.5% of 321 is what number?

Change Percents to Fractions

25% off

Regular price
$16

How much do you save by buying the clock at 25% off?

Sometimes a percent problem can be solved by changing the percent to a simplified fraction.

25% of 16 is what number?

Think:
$25\% = \frac{25}{100} = \frac{1}{4}$

$$\frac{1}{4} \times 16 = n$$

$$\frac{1}{\overset{1}{\cancel{4}}} \times \frac{\overset{4}{\cancel{16}}}{1} = n$$

$$4 = n$$

You save $4 by buying the clock on sale.

Change the percent to a fraction and multiply.

1. 50% of 80 is what number?

$50\% = \dfrac{\boxed{}}{100} = \dfrac{\boxed{}}{\boxed{}}$

$\dfrac{\boxed{}}{\boxed{}} \times \underline{\hspace{1cm}} = n$ ⌐simplified

$\underline{\hspace{1cm}} = n$

3. 10% of 60 is what number?

$10\% = \dfrac{\boxed{}}{100} = \dfrac{\boxed{}}{\boxed{}}$

$\dfrac{\boxed{}}{\boxed{}} \times \underline{\hspace{1cm}} = n$

$\underline{\hspace{1cm}} = n$

2. 20% of 40 is what number?

$20\% = \dfrac{20}{\boxed{}} = \dfrac{\boxed{}}{\boxed{}}$

$\dfrac{\boxed{}}{\boxed{}} \times \underline{\hspace{1cm}} = n$

$\underline{\hspace{1cm}} = n$

4. 25% of 32 is what number?

$25\% = \dfrac{25}{\boxed{}} = \dfrac{\boxed{}}{\boxed{}}$

$\dfrac{\boxed{}}{\boxed{}} \times \underline{\hspace{1cm}} = n$

$\underline{\hspace{1cm}} = n$

Using Fractions

For each question:

1. Change the percent to a simplified fraction.
2. Write a number sentence.
3. Solve for n.

1. 50% of 20 is what number?

$$\frac{50}{100} = \frac{1}{2}$$

↑ simplified

$$\frac{\boxed{1}}{\boxed{2}} \times 20 = n$$

_____ = n

2. 10% of 90 is what number?

$$\frac{\Box}{\Box} \times 90 = n$$

_____ = n

3. 20% of 50 is what number?

$$\frac{\Box}{\Box} \times 50 = n$$

_____ = n

4. 75% of 80 is what number?

$$\frac{\Box}{\Box} \times 80 = n$$

_____ = n

5. 25% of 120 is what number?

$$\frac{\Box}{\Box} \times 120 = n$$

_____ = n

6. 50% of 180 is what number?

$$\frac{\Box}{\Box} \times 180 = n$$

_____ = n

Using Common Equivalents

Some percents are used often. If you learn their **fractional equivalents,** you may choose to use fractions to solve many percent problems.

Learn These Common Equivalents

$$5\% = \frac{1}{20} \qquad 10\% = \frac{1}{10} \qquad 20\% = \frac{1}{5}$$

$$25\% = \frac{1}{4} \qquad 50\% = \frac{1}{2} \qquad 75\% = \frac{3}{4}$$

$$12\frac{1}{2}\% = \frac{1}{8} \qquad 33\frac{1}{3}\% = \frac{1}{3} \qquad 66\frac{2}{3}\% = \frac{2}{3}$$

Find the percent of each number.

1. 25% of 24 is what number?

$$\frac{1}{\overset{}{\underset{1}{4}}} \times \frac{\overset{6}{24}}{1} = n$$

_____ = n

4. 20% of 25 is what number?

$$\frac{\square}{\square} \times 25 = n$$

_____ = n

2. 10% of \$1,450 is what number?

$$\frac{\square}{\square} \times \$1,450 = n$$

_____ = n

5. $12\frac{1}{2}\%$ of 720 is what number?

$$\frac{\square}{\square} \times 720 = n$$

_____ = n

3. $33\frac{1}{3}\%$ of 45 is what number?

$$\frac{\square}{\square} \times 45 = n$$

_____ = n

6. 75% of 88 is what number?

$$\frac{\square}{\square} \times 88 = n$$

_____ = n

Mixed Practice

Use either a decimal or fraction to find the percent of the number. Write a number sentence for each problem and solve.

1. 50% of 90 is what number?

$$50\% \times 90 = n$$
$$\frac{1}{2} \times 90 = \underline{\hspace{2cm}}$$

2. 39% of 145 is what number?

$$.39 \times 145 = n$$
$$\underline{\hspace{2cm}} = n$$

3. 25% of 36 is what number?

4. 5% of 4,508 is what number?

5. $33\frac{1}{3}\%$ of 15 is what number?

6. 175% of 36 is what number?

7. .3% of 910 is what number?

8. 10% of 300 is what number?

9. 80% of $58.65 is what number?

10. 20% of 35 is what number?

Write a Number Sentence

1. Study the facts.
2. Find the percent and the total.
3. Set up a number sentence and solve.
4. Ask yourself, "Does the answer make sense?"

Find the part when the percent and total are given.

1. Leah earns $495, and 15% is taken out of her check for taxes. How much is taken out of her check for taxes?

   a) What is the percent? __15%__

   b) What is the total? __$495__

   c) What is the part? ___n___

2. Now write and solve a number sentence using the information given in problem 1.

$$\underset{\text{percent}}{\underline{\hspace{2cm}}}\% \ \text{of} \ \underset{\text{total}}{\underline{\hspace{2cm}}} = \underset{\text{part}}{\underline{\ n\ }}$$

$$.\underline{\hspace{1.5cm}} \times \underline{\hspace{2cm}} = \underset{\text{part}}{\underline{\ n\ }}$$

change to a decimal

$$\underline{\hspace{2cm}} = \underset{\text{part}}{\underline{\ n\ }}$$

---

3. Daniel earned $840 and saved 25%. How much did he save?

   a) What is the percent? _____

   b) What is the total? _____

   c) What is the part? _____

4. Now write and solve a number sentence using the information given in problem 3.

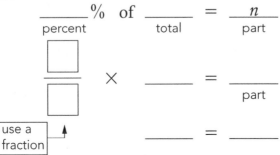

$$\underset{\text{percent}}{\underline{\hspace{2cm}}}\% \ \text{of} \ \underset{\text{total}}{\underline{\hspace{2cm}}} = \underset{\text{part}}{\underline{\ n\ }}$$

$$\frac{\square}{\square} \times \underline{\hspace{2cm}} = \underset{\text{part}}{\underline{\hspace{2cm}}}$$

use a fraction

$$\underline{\hspace{2cm}} = \underline{\hspace{2cm}}$$

Solve the Word Problems

Set up the number sentence and find the part.

1. The regular price of a sweater is $52. It is on sale for 25% off. How much is saved by buying the sweater on sale?

   __25__ % of __52__ = __n__

   _____ = _____
              part

2. Margo bought a jacket for $94.60. The sales tax was 5%. How much did she pay in sales tax?

   _____ % of _____ = _____

   _____ = _____
              part

3. Out of 600 sales, 8% were returned. How many were returned?

   _____ % of _____ = _____

   _____ = _____
              part

4. Mr. Silverman has a 140-acre farm. 75% was planted with corn. How many acres were planted with corn?

   _____ % of _____ = _____

   _____ = _____
              part

5. Mr. Shinder spends 12% of his income on recreation. If he earns $947, how much will be spent on recreation?

   _____ % of _____ = _____

   _____ = _____
              part

6. If a $435 television set is reduced in price by 30%, how much will be saved?

   _____ % of _____ = _____

   _____ = _____
              part

7. Tova wants to save $300. If she has already saved 50%, how much has she saved?

   _____ % of _____ = _____

   _____ = _____
              part

8. Mr. Lightstone wants to buy a house for $169,500. If he makes a 10% down payment, how much will he still have to pay?

   _____ % of _____ = _____

   _____ = _____
              part

Find the Part Review

Solve each problem.

1. 20% of 18 is what number?

   _____ × _____ = _____
   percent as   total   part
   a decimal

   _____ = n

2. 95% of 39 is what number?

   _____ × _____ = _____
   percent as   total   part
   a decimal

   _____ = n

3. 7% of 25 is what number?

   _____ × _____ = _____
   percent as   total   part
   a decimal

   _____ = n

4. 50% of 84 is what number?

   _____ × _____ = _____
   percent as   total   part
   a decimal

   _____ = n

5. If a $330 stereo is on sale for 20% off, how much will be saved?

   _____ × _____ = _____
   percent as   total   part
   a decimal

   _____ = n

6. Out of 440 runners, 15% finished the race. How many finished?

   _____ × _____ = _____
   percent as   total   part
   a decimal

   _____ = n

7. Riley spent $45.59. Sales tax was 8%. How much did he spend in tax?

   _____ × _____ = _____
   percent as   total   part
   a decimal

   _____ = n

8. Maria wants to save $750. She has already saved 40%. How much has she saved?

   _____ × _____ = _____
   percent as   total   part
   a decimal

   _____ = n

Compare the Numbers

Question	Picture	Fraction	Decimal	Percent
4 is what percent of 5?		$\dfrac{4}{5} \begin{array}{l}\times \\ \times\end{array} \boxed{\dfrac{20}{20}} = \dfrac{80}{100}$	.__8__ __0__	__80__ %

multiply to get a denominator of 100

	Question	Picture	Fraction	Decimal	Percent
1.	What percent of 12 is 3?	shade in	$\dfrac{\square}{\square} = \dfrac{\square}{\square} = \dfrac{\square}{100}$ (simplified)	.__ __	____%
2.	4 is what percent of 20?	shade in	$\dfrac{\square}{\square} = \dfrac{\square}{100}$	.__ __	____%
3.	What percent of 10 is 7?	shade in	$\dfrac{\square}{\square} = \dfrac{\square}{100}$	.__ __	____%
4.	4 is what percent of 8?	shade in	$\dfrac{\square}{\square} = \dfrac{\square}{\square} = \dfrac{\square}{100}$ (simplified)	.__ __	____%
5.	What percent of 50 is 37?	shade in	$\dfrac{\square}{\square} = \dfrac{\square}{100}$	.__ __	____%

Find the Percent

A $3 tip is what percent of a $20 food bill?

To find what percent one number is of another:
1. Write a fraction comparing the part to the total.
2. Rename the fraction in hundredths.
3. Write as a percent.

```
         Food Bill
_____   _____
_____   _____
_____   _____
_____   _____
_____
      Subtotal  $20.00
          Tip  $ 3.00
```

Question	Solution

3 is what percent of 20?

$$3 = n\% \times 20$$
part   total

part → $\dfrac{3}{20}$ × $\dfrac{5}{5}$ = $\dfrac{15}{100}$ = 15%
total →

multiply to get a denominator of 100

Write a fraction to compare part to total. Make the fraction a percent.

1. 8 is what percent of 10?

$$8 = n\% \times 10$$
part   total

part → $\dfrac{8}{\square}$ = $\dfrac{\square}{100}$ = _____ %
total →

3. What percent of 20 is 18?

$$n\% \times 20 = 18$$
   total   part

part → $\dfrac{\square}{\square}$ = $\dfrac{\square}{100}$ = _____ %
total →

---

2. What percent of 25 is 7?

$$n\% \times 25 = 7$$
   total   part

part → $\dfrac{\square}{\square}$ = $\dfrac{\square}{100}$ = _____ %
total →

4. 3 is what percent of 4?

$$3 = n\% \times 4$$
part   total

part → $\dfrac{\square}{\square}$ = $\dfrac{\square}{100}$ = _____ %
total →

Simplify the Fraction

If a $28 watch was marked down $7, what percent was marked off?

To find what percent one number is of another:
1. Write a fraction comparing the part to the total.
2. Simplify the fraction.
3. Change the fraction to a percent.

$$\underline{\text{Question}}$$

7 is what percent of 28?

$$\underset{\text{part}}{7} = n\% \times \underset{\text{total}}{28}$$

$$\underline{\text{Solution}}$$

$$\underset{\substack{\uparrow \\ \frac{\text{part}}{\text{total}}}}{\frac{7}{28}} \div \underset{\substack{\uparrow \\ \text{simplified}}}{\frac{7}{7}} = \frac{1}{4} \qquad \frac{1}{4} \times \frac{25}{25} = 25\%$$

multiply to get a denominator of 100

---

1. 5 is what percent of 25?

$$\underset{\text{part}}{5} = n\% \times \underset{\text{total}}{25}$$

$$\begin{array}{c}\text{part} \rightarrow \\ \text{total} \rightarrow \end{array} \frac{\Box}{\Box} = \frac{\Box}{100} = \underline{\qquad}\%$$

3. What percent of 40 is 4?

$$n\% \times \underset{\text{total}}{40} = \underset{\text{part}}{4}$$

$$\begin{array}{c}\text{part} \rightarrow \\ \text{total} \rightarrow \end{array} \frac{\Box}{\Box} = \frac{\Box}{\Box} = \frac{\Box}{100} = \underline{\qquad}\%$$

simplified

---

2. 15 is what percent of 30?

$$\underset{\text{part}}{15} = n\% \times \underset{\text{total}}{30}$$

$$\begin{array}{c}\text{part} \rightarrow \\ \text{total} \rightarrow \end{array} \frac{\Box}{\Box} = \frac{\Box}{\Box} = \frac{\Box}{100} = \underline{\qquad}\%$$

simplified

4. 24 is what percent of 32?

$$\underset{\text{part}}{24} = n\% \times \underset{\text{total}}{32}$$

$$\begin{array}{c}\text{part} \rightarrow \\ \text{total} \rightarrow \end{array} \frac{\Box}{\Box} = \frac{\Box}{\Box} = \frac{\Box}{100} = \underline{\qquad}\%$$

simplified

More Practice Finding the Percent

Write the part over the total to find the percent in each problem.

1. 8 is what percent of 40?

$$\frac{\text{part}}{\text{total}} = \frac{8}{40}$$

$$\frac{8}{40} = \frac{1}{5} = \frac{\Box}{100} = \underline{\quad}\%$$

5. What percent of 50 is 4?

2. 10 is what percent of 40?

$$\frac{\text{part}}{\text{total}} = \frac{\Box}{40}$$

6. 32 is what percent of 40?

3. What percent of 20 is 3?

7. 12 is what percent of 16?

4. 3 is what percent of 10?

8. What percent of 70 is 7?

Use Division

Sometimes we must divide to find what percent one number is of another.

1. Carry out all decimals to the hundredths place (2 decimal places).
2. Change all remainders to fractions and simplify.

A $16 calculator was marked down $6. What percent was marked off?

<u>Question</u>

6 is what percent of 16?

$$\underset{\text{part}}{6} = n\% \times \underset{\text{total}}{16}$$

Regular Price $16.00
Marked Off $6.00

<u>Solution</u>

Think:

$$\text{part} \rightarrow \frac{6}{16} = \frac{3}{8} \leftarrow \text{total}$$

$$8\overline{)3.00} \quad .37\frac{4}{8}$$
$$\underline{2\,4}$$
$$6\,0$$
$$\underline{5\,6}$$
$$4$$

$$= \underset{\text{hundredths}}{.37\frac{1}{2}} = \underset{\text{percent}}{37\frac{1}{2}\%}$$

Write the part over the total as a fraction. Then divide to find the percent.

1. 5 is what percent of 8?

$$\underset{\text{part}}{5} = n\% \times \underset{\text{total}}{8}$$

part → $\dfrac{\boxed{}}{\boxed{}}$ ← total _____%

3. 5 is what percent of 6?

$$\underset{\text{part}}{5} = n\% \times \underset{\text{total}}{6}$$

part → $\dfrac{\boxed{}}{\boxed{}}$ ← total _____%

---

2. What percent is 15 of 48?

$$n\% \times \underset{\text{total}}{48} = \underset{\text{part}}{15}$$

part → $\dfrac{\boxed{}}{\boxed{}}$ ← total $= \dfrac{\boxed{}}{\boxed{}}$ _____%

4. What percent is 30 of 32?

$$n\% \times \underset{\text{total}}{32} = \underset{\text{part}}{30}$$

part → $\dfrac{\boxed{}}{\boxed{}}$ ← total $= \dfrac{\boxed{}}{\boxed{}}$ _____%

Changing Fractions to Decimals

Divide to find what percent one number is of another number.

    Step 1: Compare the part to the total.

    Step 2: Divide and carry out all decimals to the hundredths place (2 decimal places).

    Step 3: Change all remainders to fractions and simplify.

Question	Compare	Find the Decimal	Change to Percent
2 is what percent of 16?	$\dfrac{\text{part}}{\text{total}} = \dfrac{2}{16} = \dfrac{1}{8}$	$8\overline{)1.00}$ $.12\frac{4}{8}$	$= 12\frac{1}{2}\%$

Use division to find the percents.

1. What percent of 11 is 3?

4. 21 is what percent of 24?

2. 7 is what percent of 21?

5. 2 is what percent of 18?

3. 3 is what percent of 21?

6. What percent of 21 is 14?

What Is the Percent?

To find a percent, use one of these methods.

METHOD 1	METHOD 2	METHOD 3
Write a fraction. Change to a percent.	Remember equivalent percents.	Divide. Simplify the remainder when necessary.
8 is what percent of 10?	3 is what percent of 9?	21 is what percent of 24?
$\frac{8}{10} = \frac{80}{100} = 80\%$	$\frac{3}{9} = \frac{1}{3} = 33\frac{1}{3}\%$	$\frac{21}{24} = 24)\overline{21.00}\,.87\frac{4}{8} = 87\frac{1}{2}\%$

Use one of the methods above to find the percent.

1. What percent of 24 is 6?

2. 7 is what percent of 35?

3. What percent of 21 is 7?

4. 25 is what percent of 50?

5. 16 is what percent of 24?

6. 15 is what percent of 40?

7. What percent of 7 is 5?

8. What percent of 40 is 32?

Fractions and Percents Greater Than 100%

Use the information given to find the percent in each problem.

		Fraction	Percent

1. 13 is what percent of 10?

part →

total →

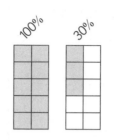

$\text{part} \to \dfrac{13}{10} = \dfrac{130}{100}$ $\underline{\quad 130 \quad}$ %

2. What percent of 5 is 9?

total   part

100%
80%

$\text{part} \to \dfrac{9}{5} \begin{smallmatrix}\times\,20\\ \\ \times\,20\end{smallmatrix} = \dfrac{\square}{100}$ _____ %

3. 6 is what percent of 4?

part   total

100%
50%

$\text{part} \to \dfrac{\square}{\square} = \dfrac{\square}{100}$ _____ %

4. What percent of 2 is 7?

total   part

100% 100% 100% 50%

$\text{part} \to \dfrac{\square}{\square} = \dfrac{\square}{100}$ _____ %

5. 32 is what percent of 20?

part   total

100%

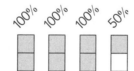
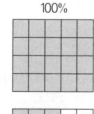

60%

$\text{part} \to \dfrac{\square}{\square} = \dfrac{\square}{100}$ _____ %

Fractions Larger Than 1

Solve the following percent problems.

1. What percent of 4 is 5?

   part —▶ $\dfrac{5}{4}$ _____%

2. 15 is what percent of 10?

   part —▶ ☐
   total —▶ ☐   _____%

3. What percent of 2 is 9?

   part —▶ ☐
   total —▶ ☐   _____%

4. 50 is what percent of 20?

   part —▶ ☐
   total —▶ ☐   _____%

5. 21 is what percent of 6?

   part —▶ ☐
   total —▶ ☐   _____%

6. What percent of 20 is 35?

   part —▶ ☐
   total —▶ ☐   _____%

7. What percent of 25 is 40?

   part —▶ ☐
   total —▶ ☐   _____%

8. 22 is what percent of 4?

   part —▶ ☐
   total —▶ ☐   _____%

Decimals and Percents
Greater Than 100%

What percent of 3 is 8?

$$\begin{array}{r} 2.66\frac{2}{3} \\ 3\overline{)8.00} \\ \underline{6} \\ 20 \\ \underline{18} \\ 20 \\ \underline{18} \\ 2 \end{array}$$

$\dfrac{\text{part}}{\text{total}}$

After dividing, change the decimal to a percent. → $266\frac{2}{3}\%$

Carry out all decimals to the hundredths place.

Use the method above to find the following percents.

1. What percent of 2 is 9?

2. 10 is what percent of 3?

3. What percent of 5 is 8?

4. What percent of 7 is 15?

5. 16 is what percent of 6?

6. What percent of 4 is 13?

Practice Your Skills

Find the percent in each problem.

1. What percent of 20 is 9?

2. 17 is what percent of 85?

3. What percent of 9 is 27?

4. What percent of 2 is 5?

5. 15 is what percent of 60?

6. What percent of 7 is 8?

7. What percent of 5 is 22?

8. 20 is what percent of 40?

9. What percent of 4 is 18?

10. 12 is what percent of 18?

Solve the Word Problems

1. Study the facts.
2. Find the part and the total given in the problem.
3. Solve for the unknown percent.
4. Ask yourself, "Does the answer make sense?"

Find the percent when the total and part are given.

1. On a test of 20 problems, Dorothy had 15 correct answers. What percent did she have correct?

   a) What is the percent? ___n___ %

   b) What is the total? __20__

   c) What is the part? __15__

   What percent of 20 is 15?

   $n\% \times \underset{\text{total}}{20} = \underset{\text{part}}{15}$

   d) part → $\dfrac{\square}{\square}$ = _____ %
      total →

2. Jack had 30 candy bars to sell. He sold 6 of them. What percent of the candy bars did he sell?

   a) What is the percent? ___n___ %

   b) What is the total? __30__

   c) What is the part? __6__

   What percent of 30 is 6?

   $n\% \times \underset{\text{total}}{30} = \underset{\text{part}}{6}$

   d) part → $\dfrac{\square}{\square}$ = _____ %
      total →

3. 16 new movies came out this month. 12 of them were adventure movies. What percent weren't adventure movies?

   a) What is the percent? _____ %

   b) What is the total? _____

   c) What is the part? _____

   What percent of 16 is 4?

   $n\% \times \underset{\text{total}}{16} = \underset{\text{part}}{4}$

   d) part → $\dfrac{\square}{\square}$ = _____ %
      total →

4. Toby has 25 grandchildren. 17 are boys. What percent of her grandchildren are girls?

   a) What is the percent? _____ %

   b) What is the total? _____

   c) What is the part? _____

   What percent of 25 is 8?

   $n\% \times \underset{\text{total}}{25} = \underset{\text{part}}{8}$

   d) part → $\dfrac{\square}{\square}$ = _____ %
      total →

More Word Problems

Set up the number sentence and find the percent.

1. A $45 sweater was marked down $15. What percent was marked off?

___n___ % of _____ = _____

part → ☐
———— = ———— %
total → ☐

2. Fred slept 6 hours out of a 24-hour period. What percent of the day did he sleep?

___n___ % of _____ = _____

part → ☐
———— = ———— %
total → ☐

3. Sam earned $400 and had $40 taken out of his check in taxes. What percent of his earnings were paid in taxes?

___n___ % of _____ = _____

part → ☐
———— = ———— %
total → ☐

4. Esther had $60 in her savings account. If she withdrew $30, what percent did she take out of her savings?

___n___ % of _____ = _____

part → ☐
———— = ———— %
total → ☐

5. It has rained 4 out of the last 6 days. What percent of days has it rained?

___n___ % of _____ = _____

part → ☐
———— = ———— %
total → ☐

6. Manny budgeted $90 out of $450 for food. What percent of his budget was for food?

___n___ % of _____ = _____

part → ☐
———— = ———— %
total → ☐

7. Ethel bought 30 apples, and 10 were spoiled. What percent of the apples were spoiled?

___n___ % of _____ = _____

part → ☐
———— = ———— %
total → ☐

8. 15 students took a test, and 12 had passing grades. What percent of the students passed the test?

___n___ % of _____ = _____

part → ☐
———— = ———— %
total → ☐

Find the Percent Review

Solve the problems.

1. What percent of 12 is 8?

    Answer: _____

2. What percent of 50 is 23?

    Answer: _____

3. 5 is what percent of 50?

    Answer: _____

4. 36 is what percent of 48?

    Answer: _____

5. 24 is what percent of 8?

    Answer: _____

6. 30 is what percent of 6?

    Answer: _____

7. What percent of 2 is 8?

    Answer: _____

8. What percent of 8 is 28?

    Answer: _____

What Is the Total?

Eva bought a pair of gloves on sale and saved $6.00. What was the regular price of the gloves?

SALE 30% OFF

Regular Price

STEP 1	STEP 2	STEP 3

STEP 1

Question:

30% of what number is $6.00?

STEP 2

Number sentence:

.30 × *n* = $6.00

percent total part
as decimal

STEP 3

Divide the part by the decimal:

$$\overset{20.}{.30\overline{)6.00}}$$

The regular price of the gloves was $20.

Change each problem to a number sentence and solve.

1. 30% of what number is 21?

.30 × *n* = 21
percent total part
as decimal

$$.30\overline{)21.00}$$

30% of _____ is 21.
total

3. 80% of what number is 16?

_____ × _____ = _____
percent   total   part
as decimal

80% of _____ is 16.
total

2. 40% of what number is 32?

_____ × _____ = _____
percent   total   part
as decimal

40% of _____ is 32.
total

4. 68% of what number is 153?

_____ × _____ = _____
percent   total   part
as decimal

68% of _____ is 153.
total

Find the Total

Jerry made a down payment of 20% on a television set. He paid $90.00 down. What was the price of the television set?

STEP 1	STEP 2	STEP 3
Question:	Number sentence:	Divide the part by the decimal:

STEP 1

Question:

20% of what number is $90?

STEP 2

Number sentence:

$$.20 \times n = \$90$$

percent   total   part
as decimal

STEP 3

Divide the part by the decimal:

$$.20\overline{)90.00}^{\,450.}$$

The price of the television was $450.

Change each problem to a number sentence and solve.

1. 75% of what number is 225?

$$.75 \times n = 225$$

percent   total   part
as decimal

$$.75\overline{)225.00}^{\,.}$$

3. 25% of what number is 15?

_____ × _____ = _____
percent   total   part
as decimal

---

2. 4% of what number is 34?

_____ × _____ = _____
percent   total   part
as decimal

4. 15% of what number is $4.65?

_____ × _____ = _____
percent   total   part
as decimal

Find the Total When the Part Is Given

Write a number sentence for each problem and solve.

1. 3% of what number is $4.23?

$$.03 \times n = 4.23$$

$$.03 \overline{)4.23}$$

2. 60% of what number is 42?

3. 6% of what number is 18?

4. 10% of what number is 6?

5. 25% of what number is 40?

6. 80% of what number is 4?

7. 15% of what number 30?

8. 90% of what number is 45?

Percent Problem Solving

1. Study the facts.
2. Find the part and the total.
3. Write a number sentence and solve.
4. Ask yourself, "Does the answer make sense?"

Find the total when the percent and part are given.

1. Judy bought a pair of shoes on sale at 20% off the regular price. If she saved $8.00, what was the regular price?

   a) What is the percent? __20__ %

   b) What is the total? __n__

   c) What is the part? __$8.00__

   Complete and solve the number sentence.

   d) _____% of _____ = _____
      percent   total   part

   e) The regular price was _____.

2. Neil saved 25% of his earnings which amounted to $150.00. How much money did he earn?

   a) What is the percent? _____%

   b) What is the total? _____

   c) What is the part? _____

   Complete and solve the number sentence.

   d) _____% of _____ = _____
      percent   total   part

   e) Neil earned _____.

Solve the Word Problems

Set up the number sentence and find the total.

1. Michigan has a 6% sales tax. If Lisa paid $30.00 in sales tax, what was the amount of her purchase?

   _____% of _____*n*_____ = $30.00

   The amount of
   her purchase was $_____.

2. Mark gave the waitress a tip of $1.80. This was 15% of the price of the meal. What was the price of the meal?

   _____% of _____ = _____

   The price of the meal was $_____.

3. 20% of Mrs. Reynolds' monthly earnings is budgeted for food. If $156.00 goes for food, what are her total monthly earnings?

   _____% of _____ = _____

   Mrs. Reynolds' total
   monthly earnings are $_____.

4. 3 students were absent, which was 10% of the class. How many students were in the class?

   _____% of _____ = _____

   There were _____ students in the class.

5. Mr. Metlis earns 40% on everything he sells. If he earned $200.00, how much did he sell?

   _____% of _____ = _____

   Mr. Metlis sold $_____.

6. The finance charge was 15%, which amounted to $39.00. What was the original price of the purchase?

   _____% of _____ = _____

   The original price
   of the purchase was $_____.

7. Marie bought a skirt on sale at 25% off. If she saved $9.75, what was the original price of the skirt?

   _____% of _____ = _____

   The original
   price of the skirt was $_____.

8. Drew lost 49 pounds, which was 20% of his original weight. What was his original weight?

   _____% of _____ = _____

   Drew's original weight was _____ pounds.

Practice Your Skills

Solve the problems.

1. 40% of what number is 34?

   Answer: _____

2. 80% of what number is 54?

   Answer: _____

3. 10% of what number is 7?

   Answer: _____

4. 68% of what number is 34?

   Answer: _____

5. Sales tax in Chicago is 8%. What is the amount of the purchase if the tax is $15.20?

   Answer: _____

6. Alfred bought a telephone at 30% off. If he saved $5.60, what was the original price of the telephone? Round to the nearest cent.

   Answer: _____

Find the Total Review

Solve each problem.

1. 25% of what number is 15?

   Answer: _____

2. 15% of what number is 45?

   Answer: _____

3. 50% of what number is 249?

   Answer: _____

4. 40% of what number is 55?

   Answer: _____

5. The delivery charge was 8%, which amounted to $14. What was the original purchase price?

   Answer: $_____

6. Etan earns 30% on everything he sells. If he earned $351, how much did he sell?

   Answer: $_____

7. Moe lost 35 pounds which was 14% of his original weight. What was his original weight?

   Answer: _____

8. 18 students were in class which was 90% of the class. How many students were absent?

   Answer: _____

Identify the Facts

Carefully read each story problem to identify the facts.

The regular price of a stereo was $420. Stella bought the stereo marked 25% off and saved $105.

What is the percent? __25__ %

What is the total? __420__ ← [all, or 100%]

What is the part? __105__

__25__% of __420__ = __105__
percent · · · total · · · part

Study the facts and then fill in each number sentence.

1. Leslie read 9 out of 12 chapters, which was 75% of the assigned lesson.

_____ of _____ = _____
percent · · total · · part

4. 3 out of 30 workers were absent. This was 10% of the workers.

_____ of _____ = _____
percent · · total · · part

2. Out of 80 students, 20 signed up for math. 25% of the students signed up for math.

_____ of _____ = _____
percent · · total · · part

5. 6 pounds of beef had 30% fat. This was 1.8 pounds of fat.

_____ of _____ = _____
percent · · total · · part

3. Julie made mistakes on only 4% of the 75 words on a typing test. She made mistakes on 3 words.

_____ of _____ = _____
percent · · total · · part

6. The basketball team won 45% of its games. The team played 20 games and won 9.

_____ of _____ = _____
percent · · total · · part

Learn the Percent Circle

The **percent circle** will help you to remember when to multiply and when to divide.

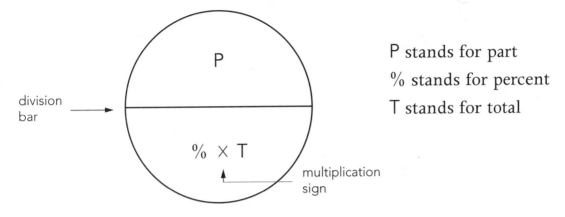

P stands for part

% stands for percent

T stands for total

---

1. Use the list to fill in the labels on the percent circle.

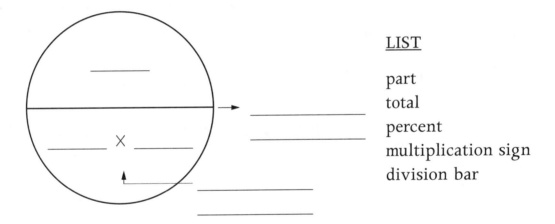

<u>LIST</u>

part
total
percent
multiplication sign
division bar

2. Build your own percent circle. Use these:

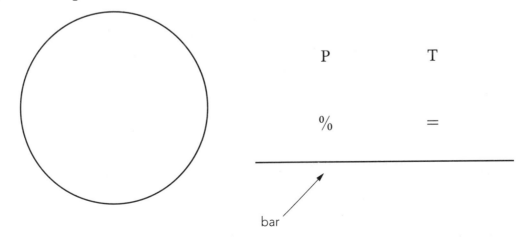

Find the Part

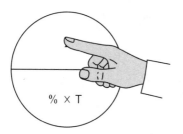

% × T

When you want to find the part:

1. Cover up the P on the circle.

2. Let the remaining symbols remind you to multiply: % × T

Example: What is 10% of 20? *(find the part)*

$$P = \% \times T \qquad\qquad .10 \times 20 = 2$$

change the percent to a decimal

2 is 10% of 20

---

1. What is 40% of 90?

   a) Cover up the _____ on the circle.

   b) The remaining symbols tell you to multiply/divide *(circle one)*.

   c) $\underline{\hspace{1cm}} \times \underline{\hspace{1cm}} = \underline{\hspace{1cm}}$
      percent   total   part

2. What is 15% of 75?

3. What is 25% of 80?

4. What is 30% of $45?

Find the Percent

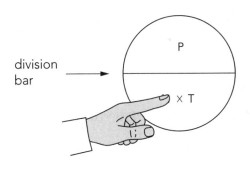

division bar

When you want to find the percent:

1. Cover up the % on the circle.

2. Let the remaining symbols remind you to divide: $\frac{P}{T}$

(The fraction bar is also a division sign.)

Example: 2 is what percent of 20? *(find the percent)*

$$\% = \frac{P}{T} = \frac{2}{20}$$

means "2 divided by 20"

$$\frac{2}{20} \times \frac{5}{5} = \frac{10}{100} = 10\%$$

2 is 10% of 20

---

1. 7 is what percent of 28?

   a) Cover up the _____ on the circle.

   b) The remaining symbols tell you to multiply/divide *(circle one)*.

   c)

   $$\frac{\square}{\square} = \frac{\square}{\square} = \underline{\quad}\%$$

   ↑ simplified

2. 13 is what percent of 65?

3. 8 is what percent of 80?

4. 7 is what percent of 175?

Find the Total

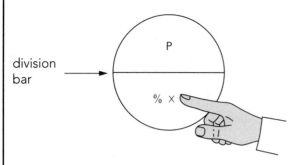

division bar →

When you want to find the total:

1. Cover up the T on the circle.

2. Let the remaining symbols remind you to divide: $\frac{P}{\%}$

Example: 2 is 10% of what number? *(find the total)*

$$T = \frac{P}{\%} = \frac{2}{.10}$$

means
"2 divided by .10" ———↑

$$.10\overline{)2.00} = 20.$$

2 is 10% of 20

---

1. 7 is 20% of what number?

   a) Cover up the _____ on the circle.

   b) The remaining symbols tell you to multiply/divide *(circle one)*.

   c) $.20\overline{)7.00}$

      7 is 20% of _____

2. 16 is 40% of what number?

3. 35 is 70% of what number?

4. 9 is 20% of what number?

Use the Circle

Use this circle to decide whether to multiply or divide.

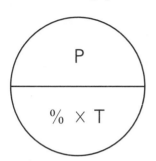

Example: 20 is what percent of 80?

To find the: part /(percent)/ total *(circle one)*

I must: multiply /(divide)*(circle one)*

Fill in: $\underset{\text{part}}{20}$ $\underset{\text{sign}}{\div}$ $\underset{\text{total}}{80}$ $= n$

$$\frac{20}{80} = \frac{1}{4} = 25\% = n$$

---

1. 16 is what percent of 48?

   a) To find the: part / percent / total *(circle one)*

   b) I must: multiply / divide *(circle one)*

   c) Fill in: $\underset{\text{part}}{\underline{\hspace{1.5cm}}}$ $\underset{\text{sign}}{\underline{\hspace{1.5cm}}}$ $\underset{\text{total}}{\underline{\hspace{1.5cm}}}$ $= n$

   d) $n = \underline{\hspace{1.5cm}}$

2. What is 75% of 60?

   a) To find the: part / percent / total *(circle one)*

   b) I must: multiply / divide *(circle one)*

   c) Fill in: $\underset{\text{part}}{\underline{\hspace{1.5cm}}}$ $\underset{\text{sign}}{\underline{\hspace{1.5cm}}}$ $\underset{\text{total}}{\underline{\hspace{1.5cm}}}$ $= n$

   d) $n = \underline{\hspace{1.5cm}}$

3. 17 is 5% of what number?

   a) To find the: part / percent / total *(circle one)*

   b) I must: multiply / divide *(circle one)*

   c) Fill in: $\underset{\text{part}}{\underline{\hspace{1.5cm}}}$ $\underset{\text{sign}}{\underline{\hspace{1.5cm}}}$ $\underset{\text{total}}{\underline{\hspace{1.5cm}}}$ $= n$

   d) $n = \underline{\hspace{1.5cm}}$

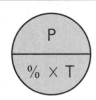

Problem-Solving Readiness

Use the percent circle to decide whether to multiply or divide. Write a number sentence and solve.

1. $33\frac{1}{3}\%$ of 45 is what number?

part

Part = $\% \times$ T

Part = _____ $\times$ _____ = _____

Think: $33\frac{1}{3}\% = \frac{1}{3}$

6. 35% of 75 is what number?

---

2. What percent of 45 is 15?

percent

Percent = $\dfrac{P}{T}$

Percent = _____ $\div$ _____ = _____

7. What percent of 16 is 40?

---

3. 5% of what number is 12?

total

Total = $\dfrac{P}{\%}$

Total = _____ $\div$ _____ = _____

change to decimal

8. 88% of 325 is what number?

---

4. 8 is what percent of 50?

9. 175% of 78 is what number?

---

5. 6.5% of 35.78 is what number?

10. 5.5% of what number is 9.35?

Identify the Facts

Fill in the facts and then complete the number sentence.

Mr. Tarnow received 180 votes. This was 60% of the 300 people who voted.

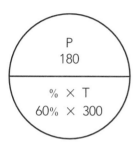

A. What is the percent? _____%

B. What is the total? __300__ ← all, or 100%

C. What is the part? _____

D. _____% of _____ = _____
     percent   total   part

Fill in the number sentence and the percent circle for each problem.

1. Benji saves 60% of $90.00 by riding the bus to work. This amounts to savings of $54.00.

_____% of _____ = _____
percent   total   part

4. Nicholas received a 10% reduction on his car insurance premium. The premium before the reduction was $350.00 so he saved $35.

_____% of _____ = _____
percent   total   part

2. Brian saved $12.00 by buying a pair of shoes on sale. The original price was $80.00 so he saved 15%.

_____% of _____ = _____
percent   total   part

5. Charles missed 8% of the problems on a test. He missed 4 out of 50 problems.

_____% of _____ = _____
percent   total   part

3. Stephanie walked to school 45 out of 180 school days. She walked to school 25% of the time.

_____% of _____ = _____
percent   total   part

6. 120 centimeters is 80% of the expected snowfall of 150 centimeters.

_____% of _____ = _____
percent   total   part

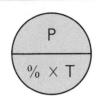

Mixed Practice

Write a number sentence for each problem. Use the percent circle to decide whether to multiply or divide. Solve each problem.

1. Zack has a weekly allowance of $20.00. He spent 30% of his weekly allowance at the roller-skating party. How much did he spend?

_____ % of _____ = _____
        total   part

Zack spent $_____ at the roller-skating party.

2. Kristina's bill at a restaurant was $40.00. She left $6.00 for the tip. What percent of the bill was the tip?

_____ % of _____ = _____
        total   part

The tip was $_____.

3. Jenna bought a mountain bike for $33\frac{1}{3}$% off the original price of $600. How much did she save?

_____ % of _____ = _____
        total   part

Jenna saved $_____.

4. If 3 out of every 5 people have at least one cold every year, what percent have at least one cold a year?

_____ % of _____ = _____
        total   part

_____ % have at least one cold a year.

5. 75% of the voters voted for Mr. Duquet. There were 8,000 people that voted. How many voted for Mr. Duquet?

_____ % of _____ = _____
        total   part

_____ people voted for Mr. Duquet.

6. The finance charge was 18%, which amounted to $27.00. What was the original price of the purchase?

_____ % of _____ = _____
        total   part

The original price of the purchase was $_____.

7. Craig earns 10% on every book he sells. If he earned $400.00, how much were the books worth?

_____ % of _____ = _____
        total   part

Craig sold $_____ worth of books.

8. The sales tax in Michigan is 6%. How much is the tax on a purchase of $45.00?

_____ % of _____ = _____
        total   part

There is $_____ tax on a purchase of $45.00.

Problem Solving

Write and solve a number sentence for each problem. Label your answers.

1. Justin bought a jacket on sale at 40% off. If he saved $34.00, what was the original price of the jacket?

2. Seth bought a new car for $16,450. He paid a sales tax of 6%. How much sales tax did he pay?

3. A taxi driver was given a $.90 tip on a $4.50 fare. What percent of the fare was the tip?

4. Nikki was given an 8% raise. This gave her $2,240 more per year. How much was her salary before the raise?

5. Victoria sold $12\frac{1}{2}\%$ of 720 candy bars. How many candy bars did she sell?
   (Hint: $12\frac{1}{2}\% = \frac{1}{8}$)

6. Nicole has $840 in a savings account that earns 2.5% interest per year. How much interest does she earn in one year?

7. 5% of Shawn's monthly earnings is budgeted for entertainment. If he spends $136.00 per month on entertainment, what are his total monthly earnings?

8. Doug was at bat 54 times and got 18 hits. What percent of his times at bat were hits?

Think It Through

Match the mathematical questions to the story problems.

_____ **1.** 2 is what percent of 8?
letter

a) Steve bought a shirt on sale at 25% off the regular price. If he saved $2.00, what was the regular price of the shirt?

_____ **2.** 25% of 8 is what number?
letter

b) If 2 cups of fruit juice are used for every 8 cups of punch, what percent of the punch is fruit juice?

_____ **3.** 2 is 25% of what number?
letter

c) Gail saves 25% of the money she earns. If she earns $8, how much does she save?

_____ **4.** 15% of 240 is what number?
letter

d) Ian has a 240-acre farm. He planted 36 acres of corn. What percent of the farm was planted in corn?

_____ **5.** 36 is what percent of 240?
letter

e) Adrienne saved $36. This was 15% of her earnings. How much did she earn?

_____ **6.** 15% of what number is 36?
letter

f) Rachel earns $240 and 15% is taken out of her check for taxes. How much is taken out for taxes?

Practice Helps

Solve each problem. Use the percent circle to help you.

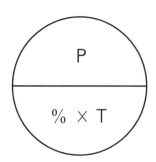

Find the part.

1. What is 35% of 80?   **Answer:** _____

2. What is 60% of 125?   **Answer:** _____

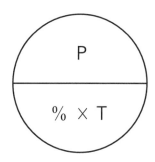

Find the percent.

3. 9 is what percent of 90?   **Answer:** _____%

4. 15 is what percent of 60?   **Answer:** _____%

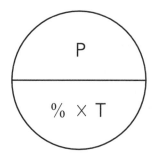

Find the total.

5. 20 is 64% of what number?   **Answer:** _____

6. 7 is 20% of what number?   **Answer:** _____

Percent Problem-Solving Review

Solve each problem.

1. What is 75% of 48?

Answer: _____

5. What is 35% of 80?

Answer: _____

2. 45 is what percent of 200?

Answer: _____%

6. 3 is what percent of 15?

Answer: _____%

3. 60 is 75% of what number?

Answer: _____

7. 120 is 48% of what number?

Answer: _____

4. Elaine has $4,550 in a CD that pays 2% interest every 6 months. How much interest will she earn in one year?

Answer: $_____

8. Debbie bought a vacation to Florida for $550. She paid 7% tax. How much tax did she pay?

Answer: $_____

Round to the Nearest Cent

Sometimes we need to round money amounts to the nearest cent.

Jeans
Regular Price
$24.95

25% of $24.95 is what number?

$$.25 \times \$24.95 = n$$

$$\$6.2375 = n$$

25% of $24.95 results in too many digits, so you must round off.

STEP 1	STEP 2	STEP 3
Circle the dollars and cents.	If the digit that follows the circled amount is **5 or more**, then **add one cent** to the circled amount.	Write the rounded amount:
$6.23⃝75 ↑ more than 5	If the digit is **less than 5**, then the circled amount **stays the same.**	$6.2375 rounds to $6.24

Round to the nearest cent.

Rounded Amount   Rounded Amount

1. ($8.84⃝8) = $8.85   6. $40.5038 = $_____
   ↑—more than 5

2. ($15.92⃝35) = $_____   7. $6.7056 = $_____
   ↑—less than 5

3. $37.857 = $_____   8. $7.005 = $_____

4. $230.149 = $_____   9. $1.1762 = $_____

5. $9.655 = $_____   10. $68.472 = $_____

Percent Off

Set up a number sentence, and find the amount of savings for each problem.

Storm Door
25% Off

Regular Price
$88

$$25\% \text{ of } \$88 = n$$

$$\underline{\hspace{2cm}} = n$$

1. The amount of savings is $_____.

Running Shoes

40% Off
Regular Price $55.50

4. The amount of savings is $_____.

Computer

10% Off
Regular Price
$899.95

2. The amount of savings is $_____.

Camcorder

15% Off
Regular Price
$499.90

5. The amount of savings is $_____.

Stereo CD Player

30% Off Regular Price
$225

3. The amount of savings is $_____.

Clock

50% Off
Regular Price
$36.50

6. The amount of savings is $_____.

Discounts

A discount is how much something has been reduced in price.

regular price − **discount** = sale price

25% OFF

If binoculars regularly sell for $96, what is the sale price?

STEP 1

Find the discount.

25% of $96 is what number?

$$\frac{1}{4} \times 96 = n$$
$$\$24 = n$$

STEP 2

Find the sale price.

Regular price − discount = sale price

$$\$96 - \$24 = \$72$$

Find the discount and sale price for each problem.

	Regular Price	Percent Off	Amount of Discount	Sale Price
1.	$89.00	20%	a) $_____	b) $_____
2.	$15.30	$33\frac{1}{3}\%$	a) $_____	b) $_____
3.	$32.00	25%	a) $_____	b) $_____
4.	$125.60	50%	a) $_____	b) $_____
5.	$234.00	30%	a) $_____	b) $_____

Find the Sale Price

To find the sale price of an item, subtract the amount of the discount from the regular price. Round off answers to the nearest cent when necessary.

1. **20%off**

Regular Price $35.99 Per Pair

  a) amount of discount $_____

  b) sale price $_____

3. **60%off**

Regular Price $99.89 each

  a) amount of discount $_____
      each

  b) sale price $_____ each

2. **25%off**

Regular Price $799

  a) amount of discount $_____

  b) sale price $_____

4. **15%off**

Regular Price $25.95

  a) amount of discount $_____

  b) sale price $_____

Discount Practice

Fill in the information.

1. The regular price of an item is $45.16. It is now selling at 25% off.

   a) What is the amount of the discount? $_____

   b) What is the sale price? $_____

2. A camera is on sale at 20% off the regular price of $325.

   a) What is the amount of the discount? $_____

   b) What is the sale price? $_____

3. Mr. Perez bought a sweater at 30% off. The regular price of the sweater was $55.10.

   a) What is the amount of the discount? $_____

   b) What is the sale price? $_____

4. A new car is selling at 5% off the regular price of $17,960.

   a) What is the amount of the discount? $_____

   b) What is the sale price? $_____

Find the discount and sale price for each problem.

	Item	Regular Price	Percent Off	Amount of Discount	Sale Price
5.	Toothbrush	$2.30	50%	a) $_____	b) $_____
6.	Battery	$59.20	20%	a) $_____	b) $_____
7.	Leather Jacket	$269.00	25%	a) $_____	b) $_____
8.	Lawn Rake	$12.40	15%	a) $_____	b) $_____
9.	Jeans	$26.88	20%	a) $_____	b) $_____
10.	Crayons	$2.60	10%	a) $_____	b) $_____

Sales Tax

State and city governments may charge a sales tax on certain items that you buy. The sales tax is given as a percent and will vary from state to state.

$$\textbf{Sales Tax} = \text{Tax Rate} \times \text{Amount of Purchase}$$

Jeffrey bought a canoe that sold for $315. The sales tax was 5%. How much sales tax did he pay?

5% of $315 is what number?

$$.05 \times \$315 = n$$

change to a decimal ⟶

$$\$15.75 = n$$

The sales tax on the canoe is $15.75.

Find the sales tax for each item. Round your answer to the nearest cent when necessary.

	Item	Amount of Purchase	Sales Tax Rate	Sales Tax
1.	Calculator	$12.95	$7\frac{1}{2}\%$	
2.	Dinner	$25.47	5%	
3.	Photo Album	$8.83	6%	
4.	Refrigerator	$496.00	5.5%	
5.	Cassette Stereo	$177.00	6.5%	

Find the Total Price

$22.95

$16.95

$65.00

$29.97

$225.00

Complete the sales slip for each item. Round your answer to the nearest cent when necessary.

1.

A-Mart		
Item	**Price**	
1 Toaster	**a)** 22.95	
1 Coffee-Maker	**b)**	
Thank You	Subtotal	**c)**
	6% Tax	**d)**
	Total	**e)**

2.

Hill's Drug Store		
Item	**Price**	
1 Camera	**a)**	
Thank You	Subtotal	**b)**
	$7\frac{1}{2}$% Tax	**c)**
	Total	**d)**

3.

Steve's Hardware		
Item	**Price**	
1 Thermos bottle	**a)**	
2 Chairs	**b)**	
Thank You	Subtotal	**c)**
	5% Tax	**d)**
	Total	**e)**

4.

Handy Helper		
Item	**Price**	
1 Coffee-Maker	**a)**	
1 Chair	**b)**	
1 Toaster	**c)**	
Thank You	Subtotal	**d)**
	$8\frac{1}{4}$% Tax	**e)**
	Total	**f)**

Sales Tax Practice

Round your answer to the nearest cent.

1. The sales tax is 6% of your $17.95 purchase.
   a) What is the sales tax? **$1.08**
      ($17.95 × .06)
   b) What is the total purchase price? $_____

2. The sales tax is 4% of your $8.99 purchase.
   a) What is the sales tax? $_____
   b) What is the total purchase price? $_____

3. The sales tax is 7% of your $125.79 purchase.
   a) What is the sales tax? $_____
   b) What is the total purchase price? $_____

4. The sales tax is 5% of your $65.98 purchase.
   a) What is the sales tax? $_____
   b) What is the total purchase price? $_____

Find the sales tax amount for each item. Then add the tax to get the purchase price.

	Amount of Purchase	Sales Tax Rate	Sales Tax	Total Purchase Price
5.	$48.79	6%	a) $_____	b) $_____
6.	$5.88	4%	a) $_____	b) $_____
7.	$12.13	5.6%	a) $_____	b) $_____
8.	$66.39	7%	a) $_____	b) $_____
9.	$23.95	5%	a) $_____	b) $_____
10.	$8.26	3%	a) $_____	b) $_____

Simple Interest in Savings

Many people who want to save money open a savings account in a bank. While the bank uses your money, it pays you **interest**—money paid for the use of your money. Three important items must be thought about when saving money at simple interest:

1. **Principal:** amount of money deposited in your savings account
2. **Rate of interest:** written as a percent
3. **Time:** figured in years

EXAMPLE

Jerry deposited $200 in a new savings account at his local bank. At 5% interest, how much will he earn in 2 years?

The simple interest rule is:

$$\text{Interest} = \text{Principal} \times \text{Rate} \times \text{Time}$$

$$\text{I} = \$200 \times 5\% \times 2$$
$$\text{I} = \$200 \times .05 \times 2$$
$$\text{I} = \$20$$

Jerry earned $20 in interest.

Find the interest for each amount of principal.

	Principal	Rate of Interest	Time	Interest
1.	$150	5%	1 year	$_____
2.	$400	6%	2 years	$_____
3.	$200	5.5%	1 year	$_____
4.	$95	8%	3 years	$_____
5.	$600	4.5%	1 year	$_____

Borrowing Money and Paying Interest

When you borrow money from a bank, there are 3 important items to think about:

1. **Principal:** amount of money borrowed
2. **Rate of interest:** written as a percent
3. **Time:** figured in years

EXAMPLE

Rick borrowed $500 for 1 year at 12% interest per year. How much must be repaid to the bank at the end of one year?

The simple interest rule is:

Interest = Principal × Rate × Time

$$I = \$500 \times 12\% \times 1$$
$$I = \$500 \times .12 \times 1$$
$$I = \$60$$

Amount to be repaid = Principal + Interest

$$= \$500 + \$60$$
$$= \$560$$

Rick must repay the bank $560 at the end of one year.

Find the interest for each amount of principal. Then find the amount to be repaid.

	Principal	Rate of Interest	Time	Interest	Amount to Be Repaid
1.	$700	9%	1 year	a) $_____	b) $_____
2.	$250	15%	2 years	a) $_____	b) $_____
3.	$15,000	8%	1 year	a) $_____	b) $_____
4.	$7,500	12%	3 years	a) $_____	b) $_____
5.	$2,000	14%	2 years	a) $_____	b) $_____

Commission

Salespeople are often paid a sum of money for selling goods. This is called a commission.

Commission = Percent × Selling Price

A real estate salesman is paid a 4% commission for every house he sells. He sold a house for $168,800. What was his commission in dollars?

SOLUTION

4% of $168,800 is what number?

$.04 \times \$168,800 = n$

$\$6,752 = n$

The salesman was paid a commission of $6,752.

Find the amount of the commission in each problem.

1. Aidan sold $250 worth of magazines and was paid a 50% commission. How much money was he paid in commissions?

   _____% of _____ = _____

   Aidan was paid $_____ in commissions.

2. Patrick's sales amounted to $82. He was paid a 15% commission. How much money was he paid in commissions?

   _____% of _____ = _____

   Patrick was paid $_____ in commissions.

3. An automobile saleswoman was paid a 6% commission. She sold a car for $19,580. What was her commission in dollars?

   _____% of _____ = _____

   Her commission was $_____.

4. Mr. Shields was paid 20% commission on sales of $5,960. How much money was he paid in commissions?

   _____% of _____ = _____

   Mr. Shields was paid $_____ in commissions.

Commission Applications

Set up a number sentence for each problem and solve. Use the percent circle to help you decide whether to multiply or divide.

1. A real estate agent was paid an 8% commission on a house valued at $93,600. How much money was the agent paid in commission?

   Answer: $_____

2. A clothing salesman earned $79.50 on sales of $1,590. What percent was he paid in commissions?

   Answer: _____%

3. Tina sold $350 worth of magazine subscriptions. If she earned $175, what percent was she paid in commissions?

   Answer: _____%

4. A real estate salesman is paid a 5% commission on sales. What was the selling price of the house if he was paid $5,270?

   Answer: $_____

5. Tom was paid a 10% commission on sales of $2,850. How much money was he paid in commissions?

   Answer: $_____

6. Rita sells books and is paid a 25% commission. If she sold $6,480 worth of books, what was she paid in commissions?

   Answer: $_____

7. An automobile salesman is paid an 8% commission. If he sold a used car for $10,942, what amount of money was he paid in commission?

   Answer: $_____

8. An insurance agent is paid a 20% commission. For sales of $850, how much money was he paid in commissions?

   Answer: $_____

Practice with Commissions

Set up a number sentence for each problem and solve. Label your answers.

1. A house sold for $75,950 and the realtor was paid a 6% commission. What was the amount of the commission?

3. Mr. Scott was paid $240.00 on sales of $1,200. What percent commission was he paid?

2. Robert was paid a 25% commission on sales. If he was paid $240.00, what were his total sales?

4. Lloyd was paid a 6% commission on sales of $95,292. How much money was he paid in commissions?

Fill in the chart below.

	Business	Rate	Total Sales	Commission
5.	Real Estate	4%	$75,300	$_____
6.	Tile and Carpeting	20%	$_____	$95.00
7.	Appliances	_____%	$300.00	$24.00
8.	Insurance	5%	$890.00	$_____
9.	Automobile	10%	$15,700	$_____
10.	Clothing	12%	$_____	$96.00

Percents and Budgets

The circle graph shows how the Rosin family budgets $2,200 each month. What amount of their budget is put aside for savings?

Family Budget

SOLUTION

15% of $2,200 is what number?

$.15 \times \$2,200 = n$

$\$330 = n$

$330 is put into savings.

Use the "Family Budget" graph to answer 1–3.

1. How much money is budgeted for clothing? _____

2. How much money is budgeted for food and rent? _____

3. How much money is budgeted for other expenses? _____

The O'Leary family is going on a vacation. The graph shows how the family budgeted $1,200 for their vacation.

Vacation Budget

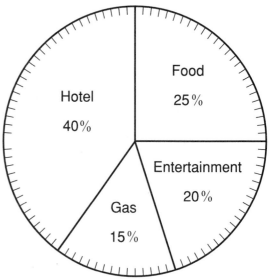

4. How much money is budgeted for food? _____

5. How much money is budgeted for gas? _____

6. How much money is budgeted for hotel? _____

7. How much money is budgeted for entertainment? _____

Percent of Increase

Sometimes percents are used to find how much an amount increased.

The price of a pair of boots increased in price from $64 to $80. What was the percent of increase?

STEP 1	STEP 2	STEP 3
Subtract to find the amount of increase.	Divide the amount of increase by the original amount.	Change the decimal to a percent.

$$\begin{array}{r} 80 \\ -\ 64 \\ \hline 16 \end{array}$$ ← original amount
16 ← increase

$$\begin{array}{r} .25 \\ 64\overline{)16.00} \end{array}$$

original amount ↗   ↖ increase

$.25 = 25\%$

The percent of increase is 25%.

Find the percent of increase.

1. A $20 watch was increased to $25.

   The percent of increase was _____%.

2. A doctor's office visit increased from $40 to $52.

   The percent of increase was _____%.

3. Lawn care service increased from $30 to $33.

   The percent of increase was _____%.

4. Gasoline increased in price from $1.20 to $1.73 per gallon.

   The percent of increase was _____%.

Percent of Decrease

Sometimes percents are used to find how much an amount decreased.

The price of a wheelbarrow decreased from $90 to $72. What was the percent of decrease?

<u>STEP 1</u>

Subtract to find the amount of decrease.

90 ← original amount
− 72
————
18 ← decrease

<u>STEP 2</u>

Divide the amount of decrease by the original amount.

$$90\overline{)18.00}^{.20}$$

original amount ↑   ↑ decrease

<u>STEP 3</u>

Change the decimal to a percent.

.20 = 20%

The percent of decrease is 20%.

Find the percent of decrease.

1. The price of the calculator decreased from $20 to $18.

   The percent of decrease was _____%.

3. The selling price of the house decreased from $125,000 to $115,000.

   The percent of decrease was _____%.

2. The price of a jacket decreased from $110 to $77.

   The percent of decrease was _____%.

4. The price of the stereo system decreased from $210 to $126.

   The percent of decrease was _____%.

Life-Skills Math Review

Solve each problem.

1. Find the sales tax. Round to the nearest cent.

    a) $22.35 purchase at 8% tax

    Tax: $_____

    b) $543.93 purchase at 7% tax

    Tax: $_____

2. Use the information from problem 1 to find the total price.

    a) Total: $_____

    b) Total: $_____

3. How much interest will be earned on $325 after 3 years at the rate of 3% per year?

    Answer: $_____

4. Aaron borrowed $550 at the rate of 14%. He has 2 years to pay back the loan. How much will he pay in interest? How much will he pay in total?

    a) Interest: $_____

    b) Total: $_____

5. A house sold for $129,900. The realtor made 7% commission. How much commission did the realtor make?

    Answer: $_____

6. A vacuum saleswoman earns 5% commission on her monthly sales. She sold $4,559 during March. How much commission did she earn?

    Answer: $_____

7. The cost of a computer table increased from $78 to $104. What was the percent of increase?

    The percent of increase was _____%.

8. The price of a barbeque decreased from $175 to $125. What was the percent of decrease?

    The percent of decrease was _____%.

Percent Application Review

1.  **25% OFF**

Regular Price
$19.88

How much do you save if you buy the blouse on sale? $_____

5. The sales tax is 5% of your $79.80 purchase.

   a) What is the sales tax? $_____

   b) What is the total purchase price? $_____

2.

Regular Price
$9.00

Save
$2.70

What percent was marked off? _____%

6. A real estate saleswoman is paid a 4% commission. If she sold a house for $138,500, what was her commission?

                    Answer: $_____

---

3.  **20% OFF**

Tony bought a CD player on sale and saved $4.80. What was the regular price of the radio? _____

7. The price of a calculator was decreased from $20 to $15. What was the percent of decrease?

                    Answer: _____%

---

4. Find each of the following:

   a) 40% of 80 is what number? _____

   b) What percent of 60 is 9? _____%

   c) 75% of what number is 90? _____

8. The regular price of an item is $38.52. The item is marked 15% off.

   a) What is the amount of the discount? $_____ (to nearest cent)

   b) What is the sale price? $_____

Solve each problem.

1. 25% of 88 is what number?

Answer: _____

2. What percent of 3 is 20?

Answer: _____%

3. 30 is what percent of 40?

Answer: _____%

4. 20% percent of what number is 72?

Answer: _____

5. Tommy earned $87.50 on sales of $1,250. What percent commission did he earn?

Answer: _____%

6. Mike bought a portable CD player that was reduced from $50 to $40. What was the percent of decrease?

Answer: _____%

7. The cost of a hotel room per night increased from $55 to $80. What was the percent of increase?

Answer: _____%

8. How much interest will be earned in a savings account at the end of 2 years if $360 was deposited at 6% interest per year?

Answer: $_____

9. Holly bought $76 hiking boots at 20% off. What was the amount of the discount?

Answer: $_____

10. Lisa bought a new softball glove for $26.50. The sales tax was 6%. What total amount did Lisa pay?

Answer: $_____

1. Find 4.5% of 300.

   Answer: _____

2. 80% of what number is 4?

   Answer: _____

3. Sandy was at bat 140 times. She hit a home run 15% of her times at bat. How many home runs did she hit?

   Answer: _____

4. Tyrone wants to save $4,500 to buy his older brother's car. So far he has saved $3,000. Tyrone has saved what percent of the total amount that he needs?

   Answer: _____

5. 400% of 16 is what number?

   Answer: _____

6. What percent of 10 is 5?

   Answer: _____

7. Joel sold $250 worth of magazines. He was paid a 20% commission. How much money did he earn on commission?

   Answer: _____

8. 30% of what number is 48?

   Answer: _____

9. What is 40% of 20?

   Answer: _____

10. 18 is what percent of 24?

   Answer: _____

11. 38 is what percent of 40?

Answer: _____

12. Matt saved $30. This amount is 20% of the total price of the books he wants to buy. What is the total price of the books?

Answer: _____

13. What percent of 300 is 6?

Answer: _____

14. Find $33\frac{1}{3}\%$ of 210.

Answer: _____

15. 75% of what number is 48?

Answer: _____

16. The sales tax rate in Steve's state is 5%. How much sales tax does Steve have to pay on a DVD player that costs $189?

Answer: _____

17. 150 is what percent of 60?

Answer: _____

18. What is $12\frac{1}{2}\%$ of 1,600?

Answer: _____

19. Cardwell is a real estate agent. He sold a house for $120,000 and received a commission of $7,200. What is his commission rate?

Answer: _____

20. Find 320% of 85.

Answer: _____

Evaluation Chart

On the following chart, circle the number of any problem you missed. The column after the problem number tells you the pages where those problems are taught. You should review the sections for any problem you missed.

Skill Area	Posttest Problem Number	Skill Section	Review Page
Percent Problems	All	7–11	12
Find the Part	1, 5, 9, 14, 18, 20	13–21	22
Find the Percent	6, 10, 11, 13, 17	7–11 23–35	12 36
Find the Total	2, 8, 15	7–11 37–42	12 43
Percent Problem Solving	3, 4, 7, 12, 16, 19	44–55	56
Life-Skills Math	All	57–72	73, 74

budget   a plan for spending money
I wrote a budget so I could save money.

commission   a percent of money paid a salesperson for selling goods
The real estate salesman earns a 6% commission on every house he sells.

discount   the reduced cost of an item
I bought my suit at a discount store to save money.

down payment   a partial payment made at the time of purchase, with the balance due later

earnings   salary, wages, or income
My earnings for this year are $23,000.00

equivalent   numbers that have the same value
$$\frac{1}{3} = .33\frac{1}{3} = 33\frac{1}{3}\%$$

finance charge   the cost of borrowing money
The finance charge on my credit card is 22%.

interest   the charge for loaned money
My bank pays 1.75% interest for my savings account.

percent   a way of expressing a number as the part of a whole; the word *percent* means *of 100*.
José answered 40 questions correctly out of 50. What percent did he answer correctly?
$$40 \div 50 = .80$$
$$= 80\%$$

premium   the amount paid for an insurance policy
The premium for my automobile insurance is $52.50 per month.

principal   a sum of money, usually in a bank or investment account
I am earning 5% on my principal investment.

reduction   to make smaller
I received a letter from my insurance company about a reduction in my insurance premium.

salary   the amount of money a person is paid in one year
My yearly salary is $23,000.00.

sales tax   tax rate × the amount of purchase; a percent charged on certain items that you buy
I pay a sales tax of $6\frac{1}{2}\%$ when I buy books.

savings account   a bank account used to save money

>   I put money in my savings account every week.

simplify (reduce)   to make the numbers in a fraction smaller without changing the value of the fraction

$$\frac{2}{4} = \frac{1}{2} \qquad \frac{4}{6} = \frac{2}{3}$$

tax rate   the percentage of money paid to the government

>   According to my paycheck, my tax rate is 23%.

tip   money left for a waiter or waitress

>   My lunch cost $6.25 and I left a $1.00 tip.

weekly allowance   money given to a person to spend each week

>   When I was young, my parents gave me a $5 per week allowance.